International Mathematics

Workbook 1

International Mathematics

Workbook 1

Andrew Sherratt

HODDER
EDUCATION
PART OF HACHETTE LIVRE UK

Hachette UK's policy is to use papers that are natural, renewable and recyclable products and made from wood grown in well-managed forests and other controlled sources. The logging and manufacturing processes are expected to conform to the environmental regulations of the country of origin.

Orders: please contact Hachette UK Distribution, Hely Hutchinson Centre, Milton Road, Didcot, Oxfordshire, OX11 7HH. Telephone: +44 (0)1235 827827. Email education@hachette.co.uk Lines are open from 9 a.m. to 5 p.m., Monday to Friday. You can also order through our website: www.hoddereducation.co.uk

© Andrew Sherratt 2008
First published in 2008 by
Hodder Education, part of Hachette Livre UK.
Carmelite House,
50 Victoria Embankment,
London EC4Y 0DZ

Impression number   11
Year   2022

Cover photo © Alan Schein Photography/Corbis
Illustrations by Charon Tec Ltd., A Macmillan Company
Typeset in 12.5/15.5pt Garamond by Charon Tec Ltd., A Macmillan Company
Printed and bound in the UK by CPI Group (UK) Ltd, Croydon, CR0 4YY

A catalogue record for this title is available from the British Library

ISBN 978 0 340 96748 5

Contents

Integers

1 Write each of these numbers using figures.

   a) five hundred and fifteen

   b) seven thousand and eighty-nine

   c) thirty-three thousand

   d) four hundred and six thousand

   e) two million, one hundred thousand

   f) three million, nine thousand and fifty-four

   g) sixty-seven billion and forty-three

   h) forty thousand, eight hundred and one

2 Write each of these numbers in words.

   a) 11

   b) 213

   c) 3867

   d) 43 291

   e) 640 552

   f) 7 849 360

   g) 404 906

   h) 15 000 020

   i) 301 200 005

   j) 12 017

3 Calculate these in your head.
(Look for short cuts to make it easier.)

   a) 31 + 16 + 9

   b) 14 + 21 + 9

   c) 14 + 6 + 9

   d) 123 + 66 + 77

   e) 67 + 52 + 33

   f) 25 + 28 + 15

   g) 7 + 25 + 13 + 75

   h) 49 + 51 + 101 + 99

   i) 28 + 22 + 41 + 59

j) $89 + 4 + 26 + 11$

k) $20 \times 819 \times 5$

l) 50×777

m) $50 \times 7303 \times 2$

n) $4 \times 2882 \times 25$

o) $25 \times 7 \times 7 \times 4$

p) 125×888

q) 25×4444

r) $11 \times 6 \times 4 \times 25$

s) $4 \times 7 \times 8 \times 5 \times 5$

t) $314 + 156 - 14$

u) $140\,000 \div 700$

4 Work out the answer to each of these, using the correct order of operations.

a) $72 - 95 + 38$

b) $48 - 69 + 47$

c) $52 + 71 - 73$

d) $26 + 45 + 37$

e) $44 - 57 - 38 + 59$

f) $77 - (54 + 35) + 43$

g) $(56 + 35) - 68$

h) $99 - (23 + 42)$

i) $92 - (59 - 43) - 49$

j) $27 - (76 - 39) + 41$

k) $83 - (94 - 58)$

l) $97 - (31 + 22) - 26$

m) $(53 - 24) - (92 - 69)$

n) $(81 - 17) - (29 + 24)$

o) $(78 - 74) \times 29$

p) $7 \times (69 - 55)$

q) $(15 + 6) \times 9$

r) $17 \times (3 + 5)$

s) $96 - 12 \times 5$

t) $13 \times 5 - 28$

u) $37 + 17 \times 6$

v) $12 \times 8 + 24$

w) $54 \div 9 + 45 \div 3$

x) $(92 - 29) \div 7$

y) $99 \div (25 - 14)$

z) $(27 + 33) \div 5$

5 Work out the answer to each of these, using the correct order of operations.

a) $96 \div (4 + 2)$

b) $37 - 84 \div 7$

c) $72 \div 8 - 5$

d) $4 + 78 \div 6$

e) $48 \div 6 + 2$

f) $162 \div (13 + 29 - 24) \times (52 - 17)$

g) $420 \div (38 - 24) \times (6 + 3)$

h) $98 \times (4 + 5) \div 14$

i) $480 \div 16 \div 15 \times 57$

j) $63 \div 7 \times 18 \div 3 \times 5$

k) $144 \div 12 \times 54 \div 6$

l) $92 - 108 \div 9 + 56 \div 8$

m) $4 \times 16 - 15 + 7 \times 23$

n) $625 \div 25 + 96 \div 8$

o) $144 \div 9 \times 5$

p) $112 \div 8 - 3 \times 4$

q) $(17 \times 9 - 33) \div (44 - 29)$

r) $96 \div [(152 - 56) \div 6] + 27$

s) $[(168 - 93) \div (37 - 12) + 57] \div 15$

t) $[(85 - 49) \div 4] \times 21$

u) $[4 \times (21 + 14) - 225 \div (76 - 51)] \times 5$

v) $[182 \div (63 - 56) + 38] \div (89 - 73)$

w) $[282 \div (12 \times 9 - 61)] + [7 \times (18 - 9) - 33]$

x) $250 \div \{29 - [160 \div (7 \times 12 - 44)]\}$

y) $17 \times (5 + 3) - \{(29 + 38) \div [180 - (55 \times 5 - 18 \times 9)]\}$

z) $\{[82 - 4 \times (4 + 3)] - 270 \div (91 - 37)\} \times [253 \div (82 - 59)]$

6 Work out each of these calculations.

   a) $16 + (-18) - (-2)$
   b) $28 + 22 + (-100)$
   c) $(-11) + 35 + (-89)$
   d) $(-18) + 10 + (-2)$
   e) $(-18) - (-10) + (-12)$
   f) $17 - 20 - (-8)$
   g) $35 - (-6) - 40$
   h) $14 + 15 - 30$
   i) $47 - (-16 - 17)$
   j) $26 - (15 - 9)$
   k) $37 + (-29) + (-18)$
   l) $-(-13) + 27 + (-20)$
   m) $[(-6) - (-16)] + (-15)$
   n) $15 - (-3) - 14 + (-24)$
   o) $[(-5) + 2] - (-13)$
   p) $(-17) + [6 + (-9)]$
   q) $[(-23) - (-18)] - [36 + (-41)]$
   r) $39 + [(-12) + (-26)] - (-99)$
   s) $31 + (-1) + (-18) + (-28)$
   t) $-[-(-21) + (-27)] - [(-66) - (-70)]$
   u) $[(-2) + (-3) + (-5)] + [10 - (-13) - (-7)] + (-1)$

7 Work out each of these calculations.

   a) $(-12) \times 5 \times 7$
   b) $15 \times (-4) \times (-3)$
   c) $(-20) \times (-5) \times 9$
   d) $(-2) \times (-3) \times (-4) \times (-5)$
   e) $16 \times 2 \times (-10)$
   f) $8 \times (-5) \times 7$
   g) $(-2) \times 5 \times (-9) \times (-7)$
   h) $(-8) \times (-3) \times 5 \times (-6)$
   i) $(-5) \times (-4) \times (-12) \times 5$
   j) $(-1) \times (-8) \times 3 \times 5$
   k) $5 \times 6 \times (-1) \times (-12)$
   l) $4 \times (-4) \times (-5) \times (-16)$

m) $10 \times (-4) \times 2 \times 5$

n) $13 \times 3 \times (-1) \times 4$

o) $9 \times 4 \times (-10) \times (-25)$

8 Work out each of these calculations.

a) $(-100) \div (-4)$

b) $(-75) \div (-25)$

c) $625 \div (-5)$

d) $56 \div (-7)$

e) $(-77) \div 11$

f) $(-49) \div 7$

g) $90 \div (-18)$

h) $36 \div (-12)$

i) $(-108) \div 2 \div (-6)$

j) $(-72) \div (-9) \div 2$

k) $45 \div (-3) \div (-5)$

l) $64 \div (-16)$

m) $(-264) \div 11 \div 8$

n) $140 \div (-7) \div 4$

o) $(-390) \div (-13) \div (-5)$

9 Work out the answers to these calculations. Remember to follow the correct order of operations.

a) $(-9) \times (-4) \div (-12)$

b) $(-56) \div [7 + (-14)]$

c) $[(-3) + (-4)] \div 7$

d) $26 \div [(-9) + (-4)]$

e) $[23 - (-5)] \div (-7)$

f) $[2 \times (-3) + (-12)] \div (-9)$

g) $[(-18) + (-3)] \div [2 \times (-4) + (-3) \times (-5)]$

h) $[(-2) \times (-5) + (-20)] \div (-10)$

i) $(-10) \times (-6) \div [(-1) - (-6)] + (-10)$

j) $[15 \times (-4) - (-6) \times 5] \div [(-8) \times (-3) - (-1) \times (-9)]$

k) $144 \div (-6) \div 3 \times (-7) \times (-5)$

l) $10\,000 \div [5 + (-25)]$

m) $[(-250) \times 4 + 280 \times (-6)] \div [(-4) + (-6)]$

n) $[660 + (-264)] \div [(-264) \div (-6)]$

o) $[(-282) \times (-5) - (-1236)] \div [5 - (-4)]$

p) $129 \div (-3) \times [(-3) + (-2)] - (-25)$

10 An aeroplane was flying at an altitude of 650 m. It descended 152 m and then ascended 834 m.

At what altitude is the aeroplane now flying?

11 At the start of a science experiment, a piece of metal had a temperature of $-24\,°C$. At the end of the experiment, the temperature of this piece of metal was $82\,°C$.

  a) What was the rise in temperature of the metal during the experiment?

  b) Calculate the temperature of the metal when it was exactly halfway between these two temperatures.

12 1 kg of ham costs $13 and 1 kg of beef costs $17.

Find the cost of 23 kg of ham and 27 kg of beef.

13 A school buys 18 footballs and 12 cricket bats for €144. One football costs €4.

Find the cost of a cricket bat.

14 The movie cinema in Flowertown has 267 seats. On Mondays, adults pay £4 each and children are free.

Last Monday the cinema was full and the owners took £540. How many children were there?

 Factors and multiples

1 Write each of these numbers as a product of its prime factors.

a) 1815

b) 3640

c) 2310

d) 264

e) 189

f) 117 800

g) 58 752

h) 61 200

i) 8008

j) 3234

k) 42 588

l) 29 645

m) 5148

n) 1925

o) 6006

p) 139 944

q) 438 867

r) 56 511

s) 937 125

t) 244 608

u) 107 712

2 Find the highest common factor (HCF) of each of these sets of numbers.

a) 12 and 30

b) 12 and 42

c) 14 and 28

d) 15 and 75

e) 16 and 40

f) 16 and 48

g) 20 and 45

h) 21 and 56

i) 24 and 64

j) 24 and 108

k) 28 and 56

l) 36 and 243

m) 45 and 42

n) 90 and 108

o) 99 and 165

p) 128 and 324

q) 45 and 63

r) 56 and 70

s) 1092 and 1170

t) 1080 and 1584

u) 1008 and 1960

v) 17 424 and 18 634

3 Find the highest common factor (HCF) of each of these sets of numbers.

a) 27, 63 and 207

b) 48, 72 and 132

c) 84, 63 and 126

d) 112, 64 and 96

e) 42, 66 and 78

f) 84, 98 and 112

g) 195, 270 and 345

h) 147, 231 and 273

i) 30, 75, 90 and 135

j) 36, 168, 144 and 252

k) 3528, 6552 and 14 616

l) 567, 441, 693, 504 and 315

m) 924, 2112, 2376 and 1188

n) 16 905, 6615 and 10 290

o) 1170, 468, 1248 and 546

p) 2640, 2112, 4752 and 3696

q) 4050, 2550, 7050 and 5550

r) 1232, 1848, 616, 968 and 792

s) 756, 1260, 2268, 3528 and 4788

4 Find the lowest common multiple (LCM) of each of these pairs of numbers.

a) 6 and 9

b) 6 and 15

c) 7 and 21

d) 8 and 12

e) 12 and 9

f) 15 and 25

g) 24 and 18

h) 30 and 25

i) 65 and 135

j) 81 and 54

k) 100 and 75

l) 120 and 135

m) 36 and 144

n) 105 and 225

o) 63 and 490

p) 160 and 400

q) 125 and 250

r) 75 and 105

s) 48 and 72

t) 30 and 24

u) 264 and 504

v) 306 and 144

w) 243 and 405

x) 261 and 435

y) 2205 and 2940

5 Find the lowest common multiple (LCM) of each of these sets of numbers.

a) 6, 9 and 15

b) 3, 12 and 16

c) 8, 9 and 12

d) 14, 18 and 21

e) 28, 44 and 68

f) 65, 135 and 175

g) 6, 8, 12 and 18

h) 6, 9, 12 and 24

i) 56, 72 and 104

j) 450, 720 and 1170

k) 300, 375 and 675

l) 378, 594 and 702

m) 252, 756 and 924

n) 36, 90, 126 and 180

o) 39, 52, 91 and 117

p) 16, 28, 44 and 68

q) 324, 756 and 972

r) 45, 90, 135 and 210

s) 84, 112, 196 and 308

t) 135, 270, 450, 630 and 675

u) 455, 520, 910, 1170 and 1560

6 Find the HCF and LCM for each of these sets of numbers.

a) 26 and 39

b) 21 and 28

c) 18 and 42

d) 336 and 224

e) 45 and 150

f) 140 and 210

g) 1008 and 1764

h) 392 and 616

i) 315 and 720

j) 84 and 189

k) 315, 525 and 1400

l) 330, 792 and 1188

m) 252, 378 and 567

n) 140, 224 and 560

o) 210, 378, 504 and 840

p) 72, 108, 144 and 162

7 Every 16 minutes, a train passes through the station at Flowerville, travelling north.

Every 45 minutes, a train passes through the station at Flowerville, travelling south.

There is a train going north and one going south both in the station together at 10 a.m.

What is the next time when two trains will be in the station together?

Unit 3   Fractions

1 Write each group of fractions in ascending order.

a) $\frac{5}{12}, \frac{7}{18}, \frac{11}{27}$

b) $\frac{13}{15}, \frac{5}{6}, \frac{37}{45}$

c) $\frac{13}{20}, \frac{11}{15}, \frac{3}{4}$

d) $\frac{7}{10}, \frac{13}{20}, \frac{2}{3}$

e) $\frac{5}{6}, \frac{7}{8}, \frac{13}{16}$

f) $\frac{9}{16}, \frac{3}{5}, \frac{11}{20}, \frac{23}{40}$

2 Write each group of fractions in descending order.

a) $\frac{2}{3}, \frac{3}{4}, \frac{4}{5}$

b) $\frac{19}{27}, \frac{13}{18}, \frac{7}{9}$

c) $\frac{14}{25}, \frac{26}{50}, \frac{53}{100}$

d) $\frac{17}{24}, \frac{5}{8}, \frac{11}{16}$

e) $\frac{3}{4}, \frac{7}{9}, \frac{13}{18}$

f) $\frac{5}{6}, \frac{13}{15}, \frac{23}{30}, \frac{7}{9}$

3 Work out the answers to these calculations. Write each answer in its lowest terms.

a) $\frac{7}{12} + \frac{11}{12}$

b) $\frac{1}{9} + \frac{5}{9}$

c) $\frac{5}{8} - \frac{3}{8}$

d) $\frac{7}{10} - \frac{3}{10}$

e) $\frac{3}{8} + \frac{1}{6}$

f) $\frac{3}{4} + \frac{7}{12}$

g) $\frac{9}{10} - \frac{11}{15}$

h) $\frac{7}{8} - \frac{5}{6}$

i) $7\frac{2}{3} + 5\frac{11}{12}$

j) $4\frac{11}{12} - 2\frac{5}{9}$

4 Work out the answers to these calculations. Write each answer in its lowest terms.

a) $2\frac{3}{4} + 3\frac{2}{3} - 2\frac{1}{2}$

b) $7\frac{5}{8} - 2\frac{1}{2} - 3\frac{3}{4}$

c) $\frac{1}{3} + \frac{7}{12} - \frac{3}{8}$

d) $\frac{2}{3} - \frac{5}{12} + \frac{3}{4}$

e) $\frac{1}{2} + \frac{3}{4} - \frac{5}{8} - \frac{3}{16}$

f) $3\frac{1}{7} + 4\frac{1}{3} - 3\frac{5}{6}$

g) $5\frac{5}{9} + 5\frac{2}{3} - 7\frac{3}{5}$

h) $7\frac{7}{8} - 2\frac{5}{6} - 2\frac{1}{4}$

5 Work out the answers to these calculations. Write each answer in its lowest terms.

a) $1\frac{1}{8} \times 1\frac{1}{3}$

b) $\frac{7}{10} \div 4\frac{1}{5}$

c) $5\frac{1}{4} \div 3\frac{1}{2}$

d) $8\frac{2}{3} \div 2\frac{1}{6}$

e) $1\frac{1}{9} \times 4\frac{1}{2}$

f) $1\frac{2}{5} \times 2\frac{6}{7}$

g) $\frac{2}{7} \times \frac{11}{90} \div \frac{1}{72} \times \frac{5}{22}$

h) $4\frac{8}{25} \times 6\frac{1}{4} \div 1\frac{5}{7}$

i) $\frac{2}{15} \div \frac{5}{12} \times 4\frac{11}{16} \div 1\frac{1}{2}$

j) $3\frac{2}{3} \times 1\frac{1}{2} \div 1\frac{4}{7} \times \frac{10}{21}$

6 Work out the answers to these calculations. Write each answer in its lowest terms.

a) $15\frac{1}{6} \div (-5)$

b) $\left(-\frac{4}{9}\right) \times \frac{3}{14}$

c) $\left(-\frac{7}{18}\right) \times \left(-\frac{9}{14}\right)$

d) $16\frac{3}{10} \times (-5)$

e) $(-8) \div \left(-\frac{4}{5}\right)$

f) $15 \times \left(-2\frac{2}{5}\right)$

g) $\frac{9}{11} \div \left(-\frac{4}{5}\right)$

h) $\left(-7\frac{1}{3}\right) \div 1\frac{5}{6}$

i) $\left(-\frac{5}{6}\right) \div \left(-\frac{3}{4}\right)$

j) $1\frac{7}{15} \div \left(-17\frac{2}{7}\right) \times 3\frac{3}{14}$

k) $\left(-2\frac{2}{5}\right) \times \frac{5}{6} \div (-13)$

l) $3\frac{3}{5} \times (-6) \div \left(-4\frac{4}{5}\right)$

7 Work out the answers to these calculations. Write each answer in its lowest terms.

Remember to follow the correct order of operations.

a) $2\frac{3}{4} + \frac{1}{2} \times 1\frac{1}{3}$

b) $1\frac{1}{3} \times \left(5\frac{1}{2} - 2\frac{3}{4}\right)$

c) $\left(4\frac{1}{2} - 2\frac{2}{3}\right) \div 3\frac{1}{3}$

d) $2\frac{1}{10} - 3\frac{1}{5} \div 2\frac{2}{3}$

e) $\left(2\frac{1}{3} - 1\frac{2}{5}\right) \div 9\frac{1}{3}$

f) $\frac{2}{3} \div \left(\frac{1}{3} + \frac{1}{4}\right)$

g) $7\frac{1}{3} \times \frac{1}{11} + 2\frac{1}{3} \div 3\frac{1}{2}$

h) $\left(2\frac{1}{2} + \frac{1}{7}\right) \div \left(3\frac{1}{2} - 2\frac{1}{13}\right)$

i) $3\frac{1}{2} \div 3\frac{1}{4} - \frac{15}{26}$

j) $1\frac{1}{11} \times 2\frac{4}{5} \times \left(\frac{5}{7} + 1\frac{1}{4}\right)$

k) $\frac{1}{4} + \left(-\frac{3}{4}\right) \times \left(-1\frac{1}{4}\right)$

l) $\left(-\frac{3}{4}\right) \times 1\frac{1}{2} + \frac{3}{4} \times \left(-2\frac{1}{2}\right)$

m) $\left[\left(-9\frac{1}{4}\right) - \left(-7\frac{3}{5}\right)\right] \div 2\frac{3}{4}$

n) $\left(\frac{1}{4} - \frac{3}{4}\right) \div \left(-\frac{1}{4}\right)$

o) $\left(\frac{7}{27} - \frac{5}{36}\right) \times (-9)$

p) $1\frac{9}{11} \times \left(\frac{3}{4} - \frac{4}{5}\right)$

8 Work out the answers to these calculations. Write each answer in its lowest terms.

Remember to follow the correct order of operations.

a) $\left(\frac{1}{4} + \frac{1}{3}\right) \div \left(1\frac{4}{5} - 1\frac{1}{3}\right)$

b) $\left(3\frac{1}{2} + \frac{5}{6}\right) \div \left(\frac{3}{4} \times 2\frac{1}{6}\right)$

c) $\left(3\frac{2}{5} - 2\frac{1}{4}\right) \div 1\frac{19}{50}$

d) $\left(1\frac{3}{4} + 1\frac{2}{3}\right) \div \frac{5}{12}$

e) $\left(3 \times 1\frac{1}{2} + \frac{1}{3}\right) \div \left(3 \times 1\frac{1}{2} - \frac{1}{3}\right)$

f) $5 \div \left[2 \div \left(5 - 3\frac{2}{3}\right)\right]$

g) $1\frac{1}{2} \div \left[1 + 1 \div \left(1 + \frac{1}{4}\right)\right]$

h) $1 \div \left[1 + 1 \div \left(1 + \frac{1}{3} \right) \right]$

i) $\left(2\frac{1}{3} \div 3\frac{1}{2} \right) \div \left(3\frac{1}{8} \times 1\frac{3}{5} \right)$

j) $\left(\frac{3}{4} - \frac{1}{3} \right) \div \left(4\frac{1}{2} - 2\frac{1}{3} \right)$

k) $\left(\frac{6}{7} \times \frac{2}{3} \div \frac{9}{14} \right) \div \left[1\frac{1}{3} \times \left(4\frac{1}{2} - 1\frac{3}{4} \right) \right]$

l) $\left[\left(3\frac{1}{2} - 1\frac{2}{3} \right) \div 3\frac{1}{3} \right] \div \left(1\frac{1}{6} + \frac{3}{10} \right)$

m) $\left[\left(1\frac{1}{2} + 2\frac{2}{3} + 3\frac{3}{4} \right) \times 3\frac{3}{5} \right] \div \left(14 - 15\frac{1}{8} \div 2\frac{1}{5} \right)$

n) $\left(8\frac{1}{3} \times \frac{1}{5} - 2\frac{1}{3} \div 3\frac{1}{2} \right) \div \left(\frac{7}{10} \times 1\frac{1}{4} + 1\frac{1}{10} - \frac{2}{3} \div \frac{5}{6} \right)$

o) $\{ 49 \div [17 - (4 \times 5 - 10)] + 4 \times 7 \} \div \left(3\frac{2}{7} - 2\frac{1}{2} \times \frac{5}{7} \right)$

9 Simone, Paulette and Amelie buy a flat together. Simone pays $\frac{3}{10}$ of the price and Paulette pays $\frac{9}{20}$.

   a) What fraction of the total price does Amelie pay?

   b) Amelie pays €70 000. What is the total price of the flat?

10 The full price of a dress is $64. In a sale, Noriko pays $\frac{3}{8}$ of the full price.

   How much does Noriko pay?

11 Mr Smith gives some money to his four children. Paul receives $\frac{1}{3}$ of the money, Gail receives $\frac{1}{4}$ and Mike receives $\frac{3}{8}$ of it. Sarah receives £90.

   How much money did Mr Smith give to his children in total?

12 Pablo uses $\frac{11}{18}$ of his garden to grow carrots. He uses $\frac{3}{7}$ of the rest of the garden to grow cabbages.

   a) What fraction of his garden is left over now?

   Pablo uses $\frac{1}{4}$ of this last piece of garden to grow peas, and the final piece to grow tomatoes.

   b) What fraction of the whole of Pablo's garden does he use to grow tomatoes?

13 Juwan has some sweets. He eats $\frac{3}{7}$ of the sweets and gives $\frac{1}{2}$ of what is left to his sister. Then he divides the sweets that remain equally between his two friends. Each friend gets 4 sweets.

How many sweets did Juwan have to start with?

14 73 pupils from Flowerville High School are travelling to the zoo. Some pupils travel on a bus with air-conditioning; and the others must travel on an old bus with no air-conditioning.

- $\frac{3}{5}$ of the pupils on the air-conditioned bus are boys.
- There are 16 girls on the bus with no air-conditioning.
- There are equal numbers of girls on each bus.

How many boys travelled on the bus with no air-conditioning?

1 Write each of these numbers in expanded form (showing the place value of each digit).

   a) 46.06

   b) 10.096

   c) 0.008 175

   d) 138.08

   e) 6543.201

   f) 9.5736

2 Write each of these numbers using decimal digits.

   a) three-tenths

   b) seven-hundredths

   c) five-thousandths

   d) four-tenths and eight-hundredths

   e) one-hundredth and three-thousandths

   f) eight and two-tenths

   g) seven and four-hundredths

   h) thirteen and nine-thousandths

   i) twenty and eight-hundredths

   j) seventy-eight, five-tenths and two-hundredths

   k) four hundred, seven-hundredths and nine-thousandths

   l) seven hundred, four-tenths and five-thousandths

   m) ninety, six-tenths and seven-thousandths

   n) eight hundred and fifteen, nine-tenths and one-ten-thousandth

3 Calculate the answer to each of these.

   a) $5.243 + 3.78 - 0.961$

   b) $3.693 - (0.712 + 0.658)$

   c) $0.36 + 4.79 + 0.32 + 2.86$

   d) $54.8 + 7.35 - 23.7$

   e) $23.28 + 4.7 - 0.53$

   f) $0.09 - 2.45 + 7.9$

   g) $10.7 - 2.94 - 6.28$

   h) $31.83 - 25.341 + 2.17$

i) $(4.35 + 1.736 - 3.58) - (3.37 - 2.19)$

j) $(4.79 - 1.81 - 3.32) + (2.13 - 1.54)$

k) $(0.402 + 8.1 + 4.8) - (0.048 + 0.81)$

l) $(1.25 + 2.7 + 5.1) - (8.1 - 0.25 - 1.7)$

4 Calculate the answer to each of these.

a) 2.94×0.083

b) $0.0481 \div 0.37$

c) 3.79×4.1

d) $422.82 \div 72.9$

e) $1193.2 \div 400$

f) 4.253×3.56

g) $0.2874 \div 0.06$

h) $0.0816 \div 0.0064$

i) 0.00083×4000

j) 4.067×600

k) $11.6 \times 4 \div (3.2 \times 0.05)$

l) $0.08 \times 0.63 \div 5.6$

m) $0.7 \times 0.6 \div 0.00168$

n) $3.7 \times 3.7 \times 3.14$

o) $14.097 \div 3.7 \div 1.5$

p) $12.6 \div (1.4 \times 4.5)$

q) $4.615 \div 0.4 \div 1.3$

r) $1.3 \times 11.3 \times 3.8 \div 279.11$

s) $3.6 \times 2.21 \div (25 \times 0.17 \times 0.04)$

t) $(0.069 \times 0.35 \times 232.2 \times 2.8) \div (0.3 \times 4.9 \times 4.3)$

 5 Calculate the answer to each of these, using the correct order of operations.

a) $30.15 \div 1.5 - 0.12 \times 0.25$

b) $6.4 - 2.88 \div 4.5 + 1.6 \times 0.85$

c) $(1.25 \times 2.7 \times 5.1) \div (8.1 \times 0.25 \times 1.7)$

d) $4.8 \times 12.5 - 25.76 \div 3.2$

e) $27.1 - (6.97 + 5.8 \times 2.25)$

f) $7.236 \div 0.18 - 14.3 \times 0.02$

g) $0.1 \div 0.25 + 1.25 \times 0.8 - 0.14$

h) $(5.36 + 7.64) \div 13 \times 0.01$

i) $36.36 \div 1.8 + 20.8 \times 3.05$

j) $28.2 - 28.2 \times 0.1 + 0.93$

k) $(0.402 \times 8.1 \times 4.8) \div (0.048 \times 0.81)$

l) $[(5.1 - 4.6) \times 0.8 + 1.04] \div 0.06$

m) $72.37 + 5.6 \times 0.35 - 0.324 \div 0.12$

n) $(0.25 + 3.68 + 0.32 + 1.75) \div 0.3 \div 12.5$

o) $2.826 \div (0.9 \times 1.57) + 63.2 \times 2.8 \div 5.53$

p) $(19.7 \times 1.8 - 0.46) \div (17.5 \times 0.16)$

q) $(4.176 \div 2.9 + 37.6 \times 3.1) \div 0.59$

r) $(3.13 + 1.6 - 0.71) \div (1.6 \times 0.25)$

s) $0.36 \div [0.36 - (2.16 \div 6 - 0.01 \div 0.25)] \times 0.7$

t) $(4.72 - 3.8 + 1.04) \div (17.61 - 17.54) - (5 - 4 \times 0.8) \div 0.15$

6 Write each of these fractions as a decimal.

a) $\dfrac{47}{100}$

b) $\dfrac{83}{1000}$

c) $\dfrac{132}{10\,000}$

d) $\dfrac{23}{20}$

e) $3\dfrac{7}{40}$

f) $\dfrac{215}{80}$

g) $53\dfrac{9}{25}$

h) $\dfrac{64}{125}$

i) $14\dfrac{17}{32}$

j) $\dfrac{108}{64}$

k) $\dfrac{43}{200}$

l) $1\dfrac{128}{125}$

m) $2\dfrac{31}{50}$

n) $\dfrac{245}{280}$

o) $\dfrac{137}{250}$

7 Write each of these decimals as a fraction in its lowest terms.

a) 0.006

b) 0.128

c) 0.075

d) 0.105

e) 0.0064

f) 0.54

g) 0.35

h) 0.86

i) 23.035

j) 7.935

k) 8.096

l) 47.48

m) 9.715

n) 13.36

o) 76.625

8 Write each of these fractions as a recurring decimal.

a) $\frac{13}{44}$ b) $\frac{59}{99}$

c) $\frac{61}{90}$ d) $\frac{7}{3}$

e) $\frac{29}{18}$ f) $\frac{11}{12}$

g) $\frac{17}{66}$ h) $\frac{5}{9}$

i) $\frac{35}{72}$ j) $\frac{4}{7}$

k) $\frac{5}{22}$ l) $\frac{29}{33}$

m) $\frac{5}{11}$ n) $\frac{47}{45}$

o) $\frac{55}{72}$ p) $\frac{19}{48}$

q) $\frac{37}{44}$ r) $\frac{100}{999}$

s) $\frac{133}{333}$ t) $\frac{1}{123}$

9 One packet of biscuits costs 67¢. How many packets can I buy with $10.05?

10 Some wood costs $3.70 per metre. How much will 9.6 m of this wood cost?

11 Shop A sells 8 toilet rolls in a pack for $2.60. Shop B sells the same toilet rolls in packs of 6 for $1.98.

Which shop sells the cheaper toilet rolls?

12 17 books cost the same as 23 CDs. One CD costs $11.90. How much does one book cost?

13 At the market I bought these items.

1.8 kg of potatoes at $2.90 per kg

2.3 kg of onions at $1.60 per kg

0.3 kg of minced beef at $12.10 per kg

7 carrots at 13¢ each.

a) What is the total amount of money that I spent at the market?

b) How much change would I receive from a $20 note?

Power numbers

1 Write each of these numbers in index form, using 2 as the base number.

a) 2   b) 16

c) 8   d) 1

e) 64   f) 32

g) 128

2 Write each of these numbers in index form, using 3 as the base number.

a) 3   b) 243

c) 27   d) 9

e) 1   f) 81

g) 729

3 Write each of these numbers in index form, using the base number shown in brackets.

a) 64 (base 4)

b) 1000 (base 10)

c) 256 (base 2)

d) 343 (base 7)

e) 216 (base 6)

f) 1 (base 13)

g) 125 (base 5)

h) 1331 (base 11)

4 Calculate these multiplications. Leave your answers in index form.

a) 2×2^7

b) $3^3 \times 3$

c) $2^4 \times 2^5$

d) $5^2 \times 5^2$

e) $10^5 \times 10^3$

f) $4^3 \times 4^6 \times 4^4$

g) $7^{10} \times 7^{21} \times 7^5$

h) $3^5 \times 3^{-7} \times 3^6 \times 3^{-2}$

i) $5^{-1} \times 5^8 \times 5^{-12}$

j) $11^{-5} \times 11^{-4} \times 11^{-2}$

k) $3^3 \times 4^4 \times 4^7 \times 3^2$

l) $9^6 \times 9^{10} \times 7^{-3}$

m) $13^{-4} \times 13^7 \times 8^5 \times 13^3 \times 8^{-9}$

n) $5^{14} \times 3^{-2} \times 6^9 \times 3^4 \times 6^{-4}$

o) $(3.1)^3 \times (3.1)^8 \times (3.2)^7 \times (3.2)^{-1}$

5 Calculate these divisions. Leave your answers in index form.

a) $4^3 \div 4^2$

b) $5^5 \div 5$

c) $4^6 \div 4^3$

d) $7^{12} \div 7^8$

e) $2^{25} \div 2^{13}$

f) $6^9 \div 6^6 \div 6^4$

g) $9^{11} \div 9^7 \div 9^2$

h) $5^7 \div 5^{-9} \div 5^8 \div 5^{-5}$

i) $7^{-3} \div 7^{10} \div 7^{-14}$

j) $13^{-7} \div 13^{-6} \div 13^{-4}$

k) $5^5 \div 6^6 \div 6^9 \div 5^4$

l) $11^8 \div 11^2 \div 9^{-5}$

m) $3^{-6} \div 3^9 \div 10^7 \div 3^5 \div 10^{-1}$

n) $7^4 \div 5^{-4} \div 8^2 \div 5^6 \div 8^{-6}$

o) $(1.7)^5 \div (1.7)^{10} \div (1.6)^9 \div (1.6)^{-3}$

6 Calculate each of these. Leave your answers in index form.

a) $8^{11} \div (8^8 \times 8^6)$

b) $2^2 \times (2^9 \div 2^4)$

c) $(7^9 \times 7^{-2}) \div (7^1 \div 7^{-7})$

d) $9^{-5} \div (9^2 \div 9^{-12})$

e) $(3^{-9} \div 3^{-8}) \times (3^{-6} \div 3^7)$

f) $7^7 \div 8^8 \times 8^2 \div (7^6 \div 7^{-3})$

g) $13^{-1} \div 13^4 \div 2^{-7} \times 13^3$

h) $5^{-8} \times 5^{11} \times 2^9 \div 5^7 \div 2^{-3}$

i) $9^6 \div 7^{-6} \times 10^4 \div 7^8 \times 10^{-8}$

j) $(2.1)^7 \div (2.1)^2 \times (2.7)^{11} \div (2.7)^{-5} \times (2.1)^{-4}$

7 Calculate each of these. Leave your answers in index form.

a) $\dfrac{2^6 \times 2^3}{2^2}$

b) $\dfrac{3^7 \times 3^3}{3^5}$

c) $\dfrac{4^{15} \times 4^6}{4^{12}}$

d) $\dfrac{5^5 \times 5^{12}}{5^9}$

e) $\dfrac{6^8 \times 6^2 \times 6}{6^5}$

f) $\dfrac{7^{10} \times 7^3 \times 7^1}{7^7}$

g) $\dfrac{8^6 \times 8^7}{8^4 \times 8^5}$

h) $\dfrac{9^5 \times 9^6}{9^2 \times 9^4}$

i) $\dfrac{11^8 \times 11^2 \times 11^4}{11^{12} \times 11}$

j) $\dfrac{12^7 \times 12^9 \times 12}{12^{11} \times 12^5}$

8 Calculate each of these. Leave your answers in index form.

a) $\dfrac{2^2 \times 9^3}{2^2 \times 9^2}$

b) $\dfrac{3^6 \times 8^7}{8^3 \times 3^4}$

c) $\dfrac{4^6 \times 7^{11}}{7^4 \times 4^5}$

d) $\dfrac{5^8 \times 6^9}{6^5 \times 5^4}$

e) $\dfrac{3^5 \times 2^{10}}{3^2 \times 2^4}$

f) $\dfrac{5^4 \times 4^7}{4^4 \times 5^3}$

g) $\dfrac{8^4 \times 7^7}{8^3 \times 7^6}$

h) $\dfrac{10^{12} \times 9^5}{9^4 \times 10^7}$

i) $\dfrac{2^{14} \times 4^8 \times 4^{-3}}{4^4 \times 2^9}$

j) $\dfrac{3^4 \times 5^{15} \times 5^{-6}}{5^3 \times 3}$

9 Write each of these numbers in standard form.

a) 400

b) 5000

c) 53

d) 780

e) 94 000

f) 143 000

g) 822 822

h) 76.3

i) 14.59

j) 63.7

k) 489.5

l) 3 454 454

m) 35.008

n) 7823

o) 74 392

p) 6 923 746

q) 47 000 000

r) 523.9

s) 8274

10 Write each of these numbers in standard form.

a) 0.04

b) 0.0007

c) 0.005 83

d) 0.0902

e) 0.000 069

f) 0.000 000 045

g) 0.0048

h) 0.0172

i) 0.000 003 7

j) 0.006 009

11 Write each of these numbers in standard form.

a) 33×10

b) 527×10^{-1}

c) 0.7×10

d) 0.024×10

e) 99.9×10^{6}

f) 61.83×10^{2}

g) 65.3×10^{-2}

h) 678.5×10^{3}

i) 36.7×10^{4}

j) 325×10^{-7}

k) 86.7×10^{-4}

l) 12.4×10^{-3}

m) 4823×10^{-11}

12 Calculate these. Give your answers in standard form.

a) $9 \times (23 \times 10^{6})$

b) $4 \times (5.72 \times 10^{-7})$

c) $34 \times (5.2 \times 10^{-2})$

d) $0.5 \times (16.7 \times 10^{-9})$

e) $3 \times (85.7 \times 10^{5})$

f) $3.2 \times (5 \times 10^{7})$

g) $7 \times (43.6 \times 10^{10})$

h) $3.6 \times (5.5 \times 10^{12})$

i) $5 \times (2.87 \times 10^{-2})$

j) $(87 \times 10^{12}) \times (56 \times 10^{-8})$

k) $(7.9 \times 10^{10}) \times (3.6 \times 10^{7})$

l) $(49 \times 10^{-21}) \times (66 \times 10^{7})$

m) $(75 \times 10^{4}) \times (2.4 \times 10^{-1})$

n) $(4.3 \times 10^{4}) \times (5 \times 10^{9})$

o) $(231.4 \times 10^{-7}) \div (3.56 \times 10^{-12})$

p) $(8.4 \times 10^{9}) \div (20 \times 10^{-4})$

q) $(1.651 \times 10^{22}) \div (2.54 \times 10^{10})$

r) $(81 \times 10^{-6}) \div (3.24 \times 10^{5})$

13 In 2006, the total number of people in a country was sixty-seven million, eight hundred and forty-five thousand. Three hundred and sixty-two thousand were children under the age of 16.

How many people in this country were older than 16? Write your answer in standard form.

14 In 2005, 536 000 aeroplanes used Heathrow airport in London.

If each aeroplane carried 287 people, how many people used Heathrow airport in 2005? Write your answer in standard form.

1 Write each of these Roman numerals in modern decimal numbers.

  a) LVII
  b) LXXXI
  c) XXIII
  d) VII
  e) XCIV
  f) XLV
  g) XXXVI
  h) XVIII
  i) XXXIX
  j) LXXXVIII
  k) LXVI
  l) XLIV
  m) XXIX
  n) LI
  o) XIV
  p) LXXII
  q) XCV
  r) XXXIII
  s) XII
  t) IX

2 Write each of these Roman numerals in modern decimal numbers.

  a) CDXLIX
  b) CCCLXXIII
  c) CXIV
  d) CCVI
  e) CDLXXXII
  f) CCCLXIX
  g) CLXXII
  h) CCV
  i) CCCXXVI
  j) CXI

k) MCDLVIII

l) MMMCMV

m) MMXXXIII

n) MMMXCVI

o) MMDLVII

p) MCMLXV

q) MMMDCCCLXXXI

r) MDCCLXXIV

s) MMDCCCXCVIII

t) MMMDXXII

3 Write each of these numbers using Roman numerals.
(Use upper case.)

a) 79		**b)** 13	
c) 46		**d)** 11	
e) 26		**f)** 67	
g) 58		**h)** 30	
i) 54		**j)** 22	
k) 77		**l)** 38	
m) 41		**n)** 75	
o) 34		**p)** 96	
q) 17		**r)** 55	
s) 34		**t)** 19	

4 Write each of these numbers using Roman numerals.
(Use upper case.)

a) 661	**b)** 595
c) 336	**d)** 428
e) 604	**f)** 581
g) 394	**h)** 227
i) 548	**j)** 372
k) 2569	**l)** 1016
m) 3144	**n)** 1107
o) 3668	**p)** 2176
q) 1992	**r)** 2885
s) 3909	**t)** 1633

Approximation and estimation

1 Round each of these numbers

   i) to the nearest 10

  ii) to the nearest 100

 iii) to the nearest 1000.

  a) 41 864   **b)** 553 356

  c) 671 097   **d)** 1 156 115

  e) 3 580 074   **f)** 771 137

  g) 938 730   **h)** 39 049

   i) 899 994   **j)** 3 786 666

  k) 66 837   **l)** 491 125

  m) 2 279 983   **n)** 87 747 813

  o) 505

2 Round each of these as described.

  a) 567 g to the nearest 10 g

  b) 833 g to the nearest 100 g

  c) 4.38 mm to the nearest millimetre

  d) 239 567 people to the nearest 1000 people

  e) 29.3 to the nearest whole number

  f) 42.367 m to the nearest 10 m

  g) 23.46 cm to the nearest 0.1 cm

  h) 5436 pupils to the nearest 100 pupils

  i) 956 km to the nearest 10 km

  j) 33.69 kg to the nearest 0.1 kg

3 Write each of these numbers correct to one decimal place.

  a) 7.34   **b)** 30.95

  c) 45.39   **d)** 37.98

  e) 57.17   **f)** 249.09

  g) 47.249   **h)** 3.508

  i) 29.6186   **j)** 227.7651

  k) 25.3188   **l)** 3.0942

4 Write each of these numbers correct to two decimal places.

a) 5.799

b) 37.9038

c) 0.056

d) 25.048

e) 2.523

f) 3.4462

g) 21.1979

h) 9.0985

i) 5.1872

j) 23.6047

k) 7.9286

l) 239.349 51

5 Round each of these numbers correct to three decimal places.

a) 1.148 75

b) 9.5564

c) 21.6174

d) 8.5516

e) 22.7498

f) 7.9393

g) 33.7576

h) 43.139 51

i) 24.561 94

j) 5.936 62

k) 1.696 94

l) 0.009 286 4

6 Round each of these decimals to the number of decimal places given in brackets.

a) 6.886 592 (4 d.p.)

b) 1.199 360 2 (2 d.p.)

c) 0.336 651 3 (3 d.p.)

d) 0.000 77 (4 d.p.)

e) 0.0488 (2 d.p.)

f) 48.239 47 (3 d.p.)

g) 1.826 584 7 (1 d.p.)

h) 0.003 975 (4 d.p.)

i) 25.997 483 (1 d.p.)

j) 834.687 25 (2 d.p.)

k) 1.422 196 75 (3 d.p.)

l) 99.598 964 (2 d.p.)

7 Write down the number of significant figures in each of these numbers.

a) 1.67   b) 50.60

c) 0.60   d) 87.08

e) 0.08   f) 27.004

g) 0.000 89   h) 0.97

i) 50 097   j) 300 000

k) 26.0   l) 40 867

m) 5.03   n) 74

o) 40.09   p) 59.30

q) 7 000 000   r) 0.0394

s) 0.85   t) 0.000 76

8 Write each of these numbers correct to three significant figures.

a) 5.149   b) 3.675 61

c) 7198   d) 0.800 49

e) 0.005 998 3   f) 0.087 963

g) 29.09   h) 4.1165

i) 0.009 276 5   j) 0.043 67

9 Write each of these numbers correct to four significant figures.

a) 251.19   b) 38.9038

c) 0.090 504 6   d) 700 969

e) 1.946 46   f) 6004.694

g) 29.2467   h) 350.0391

i) 50 003   j) 3.0068

10 Round each of these numbers to the number of significant figures written in brackets.

a) 4.095 (2 s.f.)

b) 2.594 67 (4 s.f.)

c) 0.004 47 (1 s.f.)

d) 27.058 (1 s.f.)

e) 4.242 68 (2 s.f.)

f) 0.683 49 (2 s.f.)

g) 0.067 89 (3 s.f.)

h) 328.007 (5 s.f.)

i) 4.699 (3 s.f.)

j) 26.8048 (4 s.f.)

k) 0.004 67 (2 s.f.)

l) 6.97 (2 s.f.)

m) 0.0462 (2 s.f.)

n) 28.96 (3 s.f.)

o) 9.464 (2 s.f.)

p) 230.509 (5 s.f.)

q) 5.937 (3 s.f.)

r) 23.097 (2 s.f.)

s) 0.059 83 (2 s.f.)

t) 0.090 47 (2 s.f.)

u) 0.020 20 (3 s.f.)

v) 0.204 38 (3 s.f.)

w) 2 975 962 (2 s.f.)

x) 5.0483 (2 s.f.)

y) 0.069 87 (1 s.f.)

11 Work out each of these calculations and

i) write the answer correct to three decimal places

ii) write the answer correct to three significant figures.

a) $\dfrac{28.58 \times 7.98}{6.72 - 4.63}$

b) $\dfrac{(7.63)^2 \times (4.96)^2}{211.4}$

c) $\dfrac{2.854 + 6.4}{(12.82)^2}$

d) $\dfrac{6.394 \times 17.58}{0.0468 \times 53.875}$

e) $\dfrac{2.44 \times 0.938 \times 3.348}{4.685 - 0.379}$

f) $\dfrac{31.849 \times (5.827)^2}{0.788 \times 3.8201}$

12 Work out each of these calculations and

    i) write the answer correct to four decimal places

    ii) write the answer correct to four significant figures.

a) $\dfrac{121.84 - 14.38 \times 5.822}{99.88 + 67.289}$

b) $\dfrac{25.811}{(19.124 - 16.188) \times 20.1614}$

c) $\dfrac{9.988 \div 44.217}{38.214 - 50.502 \times 0.6791}$

d) $\left(\dfrac{33.52 - 11.581}{8.329}\right) \times \left(\dfrac{5.8214 \times 22.594}{9.570\,2}\right)$

e) $\left(\dfrac{93.534 \times 15.724}{13.589}\right) \div \left(\dfrac{5.55 \times 3.327544}{9.964}\right)$

13 Michael has \$8.00. A pen costs 89¢.

How many pens can Michael buy?

14 79 pupils are visiting the zoo on a school trip. Each minibus can carry 25 pupils.

How many minibuses will the school need to carry all the pupils to the zoo?

15 For Parents' Day, some pupils want to display their pictures on the classroom wall. The classroom wall measures 487 cm and each picture is 31 cm wide.

How many pictures can they display on the wall?

16 A shop has 500 balloons. This shop sells 36 balloons in a bag.

How many bags of balloons can they sell?

17 Seven school friends decide to order some fried chicken. Each friend wants at least two pieces of chicken. Each box of fried chicken contains six pieces.

How many boxes must the friends buy?

18 Estimate the answer to each of these calculations by rounding correct to one significant figure before calculating.

    a) 5.86×302.98

    b) 5882.4×334.85

    c) $35\,050 \times 9.967$

    d) 4872×690

    e) $6949 \div 0.049\,866$

    f) 7.295×0.7789

    g) $42\,991 \times 6911$

    h) $223.21 \div 0.3988$

    i) $7.12 \div 0.0442$

    j) $6901 \div 43\,921$

    k) $4934 \div 0.004\,78$

    l) $8.97 \div 3.03$

19 For each of these calculations, estimate the answer by rounding to one significant figure.

Do NOT work out the exact answers.

Write each answer correct to one significant figure.

    a) $\dfrac{68.37 \times 6.218}{4.928}$

    b) $\dfrac{27.13 \times 0.0442}{0.088\,32}$

    c) $\dfrac{98.93}{21.824 \times 0.039\,26}$

    d) $\dfrac{10.145 \times 0.3207}{0.0391 \times 35.889}$

    e) $\dfrac{3816 \times (0.623)^2}{0.005\,998}$

    f) $\dfrac{4.422 \times 0.038\,27}{892 \times 0.0342}$

    g) $\dfrac{28.23 \times 5.278}{2.906}$

    h) $\dfrac{52.52}{21.03 \times 0.029\,76}$

i) $\dfrac{4.9 \times 68 \times 0.491}{31.239}$

j) $\dfrac{0.392 \times 0.008\,35}{21.249 \times 0.0887}$

k) $\dfrac{(49.86)^2 \times 5.2413}{13.497}$

l) $\dfrac{6.8053 \times 9.034}{6.923 - 3.344}$

20 There are 44 classrooms at Flowerville School. Each classroom can hold 38 pupils.

Estimate how many pupils can study at this school.

21 Pascal's car can travel 10.67 km on 1 litre of petrol. His petrol tank holds 58.43 litres of petrol.

Estimate how far Pascal can drive on one tank of petrol.

22 Mr Lu's annual salary is $34 799.63.

a) Estimate his monthly salary.

b) On his tax forms, Mr Lu must round his salary to the nearest dollar.

What should he write?

23 Mrs Singh is buying a new dining table and six chairs. The table costs $475 and each chair costs $119.

Estimate how much money Mrs Singh will spend.

Unit 8

Measures and measurement

1 Convert (change) each of these times to the units given in brackets.

   a) 12 days (hours)

   b) 5 hours (minutes)

   c) 11.5 hours (minutes)

   d) 390 seconds (minutes)

   e) 102 hours (days)

   f) 585 minutes (hours)

   g) 260 minutes (hours)

   h) 17 minutes (seconds)

   i) 80 hours (days)

   j) 26 100 seconds (hours)

   k) $4\frac{1}{3}$ days (minutes)

   l) $6\frac{5}{6}$ hours (seconds)

2 Add up these times.

   a) 3 h 11 min + 6 h 17 min

   b) 7 h 46 min + 2 h 9 min

   c) 8 h 39 min + 3 h 42 min

   d) 17 h 17 min + 15 h 47 min

   e) 11 h 27 min + 4 h 52 min

   f) 9 h 36 min + 9 h 39 min

3 Subtract these times.

   a) 9 h 47 min − 4 h 22 min

   b) 7 h 23 min − 4 h 14 min

   c) 3 h 15 min − 1 h 26 min

   d) 22 h 7 min − 19 h 12 min

   e) 15 h 23 min − 7 h 55 min

   f) 8 h 33 min − 2 h 37 min

4 Rewrite each of these times using the 24-hour clock.

a) 10.13 a.m.   **b)** 9.46 a.m.

c) 12.27 a.m.   **d)** 1.38 a.m.

e) 6.19 a.m.   **f)** 2.37 p.m.

g) 4.55 p.m.   **h)** 7.16 p.m.

i) 10.21 p.m.   **j)** 8.40 p.m.

5 Rewrite each of these times using the 12-hour clock.

a) 01 10   **b)** 05 50

c) 00 10   **d)** 10 40

e) 12 20   **f)** 18 06

g) 15 15   **h)** 19 09

i) 21 21   **j)** 23 59

6 Calculate how much time is between each of these two times (in the same day).

a) 05 00 and 11 00

b) 08 30 and 21 00

c) 15 45 and 18 30

d) 14 40 and 22 20

e) 12 27 and 20 19

f) 6.00 a.m. and 4.00 p.m.

g) 7.30 a.m. and 7.45 p.m.

h) 12.48 a.m. and 12.13 p.m.

i) 9.37 p.m. and 11.22 p.m.

7 In each pair of times below, the first time is 'today' and the second time is 'tomorrow'.

Calculate how much time is between each of the two times.

a) 17 00 and 07 00

b) 14 30 and 09 45

c) 06 15 and 03 45

d) 19 37 and 07 17

e) 00 43 and 12 06

f) 9.00 p.m. and 6.00 a.m.

g) 4.30 p.m. and 11.15 a.m.

h) 10.20 a.m. and 2.49 a.m.

i) 11.09 p.m. and 12.27 a.m.

8 On Monday, a train leaves Bristol station at 21 53 and arrives in Glasgow at 04 36 on Tuesday.

How long does this journey take?

9 Valentin earns $5.40 in 1 hour. Calculate how much he earns in

  a) 40 minutes

  b) $7\frac{3}{4}$ hours.

10 Mrs Bocanegra works in a shop for $8\frac{1}{2}$ hours each day from Monday to Friday.

She starts work at 10.15 a.m. and she earns $289 per week.

  a) At what time does Mrs Bocanegra finish work each day?

  b) How much does she earn in one hour?

11 Mr Sharma travels for 49 minutes to work each morning and 1 hour 7 minutes back home each evening. He arrives at work at 8.27 a.m. and leaves work at 5.05 p.m. Calculate

  a) the time Mr Sharma leaves home each morning

  b) the time Mr Sharma arrives at home each evening

  c) how long Mr Sharma spends at work each day.

12 Marija walks for 37 minutes to get to school each day from Monday to Friday.

Calculate how many hours and minutes altogether Marija walks to and from school in one week.

13 Sven works six days each week in a factory. He works for $4\frac{3}{4}$ hours each morning and $3\frac{1}{2}$ hours each afternoon. His lunch break is from 12.15 p.m. until 1 p.m. each day. Calculate

  a) at what time Sven starts work each day

  b) at what time Sven finishes work each day

  c) how many hours and minutes Sven is at work each week (including his lunch breaks).

14 Convert each of these lengths to the unit given in brackets.

    a) 5 m (cm)   **b)** 6 km (m)

    c) 4 m (mm)   **d)** 8123 cm (mm)

    e) 72 mm (cm)   **f)** 6724 mm (m)

    g) 32 cm (mm)   **h)** 6.1 km (m)

    i) 7.35 m (mm)   **j)** 804 m (km)

    k) 614 cm (km)   **l)** 147 mm (m)

   m) 71.6 m (cm)   **n)** 8.59 cm (mm)

    o) 58.4 m (cm)   **p)** 67 m (mm)

    q) 93 km (cm)   **r)** 0.0378 km (mm)

    s) $35 \frac{7}{10}$ km (m)   **t)** $52 \frac{3}{4}$ cm (mm)

    u) $17 \frac{4}{5}$ m (mm)   **v)** $23 \frac{7}{8}$ km (cm)

15 A snail crawled 730 cm away from a bush in the morning and 9.8 m further away in the afternoon.

At night, the snail crawled 9352 mm back towards the bush. How far from the bush is the snail now? Write the answer in metres.

16 Raisa can walk 4820 m in one hour. She starts walking at 9.45 a.m.

How far has she walked by 1.09 p.m.? Write the answer in kilometres.

17 Convert each of these masses to the unit given in brackets.

    a) 7.9 kg (g)   **b)** 3762 kg (t)

    c) $6 \frac{3}{8}$ kg (g)   **d)** 853 t (kg)

    e) 329 kg (t)   **f)** 4.46 t (kg)

    g) $9 \frac{7}{8}$ t (kg)   **h)** 74 g (kg)

    i) 3026 mg (g)   **j)** 5.638 g (mg)

    k) 863 g (t)   **l)** 0.000 56 t (g)

18 Four bags of flour and three bags of sugar have a total mass of 3.9 kg. One bag of sugar has a mass of 460 g.

What is the mass of one bag of flour?

19 One carrot contains 15.7 mg of vitamin A.

What is the mass of vitamin A contained in 200 carrots?
Write the answer in grams.

20 200 oranges contain a total of 4.68 g of vitamin C.

What is the mass of vitamin C in one orange?
Write the answer in milligrams.

21 A bus has a mass of 2.74 tonnes. Mary weighs 45.3 kg, Luka weighs 62.7 kg and Mendel weighs 64.8 kg. The spare tyre weighs 37.4 kg.

What is the new mass of the bus if the three children get on and the spare tyre is taken off the bus? Write the answer in tonnes.

22 Convert each of these volumes to the unit given in brackets.

a) 93 534 ml (litre)   b) 84 litres (ml)

c) 3.796 litres (ml)   d) 26.915 litres (ml)

e) 3978 ml (litre)   f) 9557 ml (litre)

g) $17\frac{5}{8}$ litre (ml)   h) $154\frac{3}{4}$ litre (ml)

i) 47.3 ml (litre)

23 John has $2\frac{1}{2}$ litres of milk. He drinks 0.325 litres of the milk.

How many millilitres of milk are left over?

24 To make some grey paint Maria mixes half a litre of white paint with 15 millilitres of black paint.

How much grey paint does Maria have altogether?
Give the answer in litres.

1 Choose the correct algebraic expression for each of the sentences.

    a) 5 times a number, decreased by 8

    b) 5 times the sum of a number and 8

    c) 5 more than 8 times a number

    d) 8 times the sum of a number and 5

    e) twice the sum of 5 times a number and 8

    f) 2 more than 5 eighths of a number

    g) 8 times the difference of twice a number and 5

    A $8(x + 5)$

    B $8(2x - 5)$

    C $8x + 5$

    D $2(5x + 8)$

    E $5x - 8$

    F $5(x + 8)$

    G $\frac{5}{8}x + 2$

2 Write each of these sentences as an algebraic expression.

    a) nine times c plus five times d

    b) six times j squared, minus twice g squared

    c) five times the square of a, plus four times b squared

    d) the product of seven and x, plus y multiplied by one third of z

    e) seventeen subtracted from twice the number s

    f) six times the number c decreased by eleven

    g) four times the number d divided by the sum of eight and e

    h) eight times the sum of f and four times g, minus the quotient of h and twice k

    i) twice the sum of e and thirty, reduced by forty

    j) two-thirds the sum of n and three-sevenths of p

    k) the product of a and b decreased by twice the difference of c and d

l) five more than the sum of y and six

m) the quotient of five more than twice s, and s

n) five increased by one-half of the sum of t and three

o) the quotient of fifteen, and the sum of m and twelve

3 Evaluate each of these expressions when
$a = 1$, $b = 2$, $c = 3$, $w = 0$, $x = 10$ and $y = 6$.

a) xy

b) $b + cy$

c) $x - ac$

d) $7b + 4c$

e) $8x - 3y$

f) $ax + by$

g) $2x(b + c)$

h) $\dfrac{x + y}{c - a}$

i) $\dfrac{xy}{x + b}$

j) $\dfrac{wa}{b}$

k) $(x - y)(y - w)$

l) $c(y + c)(y - c)$

m) $\dfrac{3x}{b}\,abc$

n) $8bc - (w + x + y)$

o) $\dfrac{x - b}{y + b}$

4 Evaluate each of these expressions when
$x = 2$, $y = -3$ and $z = 4$.

a) $\dfrac{7x + 8y + 2z}{3z - 2y - 4x}$

b) $\dfrac{5x + 3yz - z^2}{2xz - 4y}$

c) $\dfrac{3x^3 + 2y^2 - 4z}{2x + 4y}$

d) $\dfrac{5xz - 2y^2}{2xy}$

e) $\dfrac{2z - 4y + 5xy}{(z + x)(z - x)}$

f) $\dfrac{\dfrac{2}{y} + \dfrac{z}{x}}{\dfrac{z}{y} - \dfrac{x}{z}}$

g) $\dfrac{\dfrac{x}{y} \times \dfrac{z}{y}}{\dfrac{x}{y} - \dfrac{x}{z}}$

h) $\dfrac{\dfrac{1}{x} - \dfrac{2}{y}}{\dfrac{y}{x} \div \dfrac{1}{z}}$

5 Simplify each of these expressions.

a) $-2y + 7y + 4$

b) $5x + 7 + x - 9x$

c) $-8y - 2y - 4 + 4y$

d) $6x - (-3x) + x - 6$

e) $4x + 2y + 4x - 5y$

f) $6x + 8y - 3 - 7y$

g) $-6x - 2y + 8 + 5x - 1$

h) $9 - 3x - (-8y) + 9x - y$

i) $x - 4y - 12 - 5y + 8y$

j) $3x + 7 - 7y + 2x - 3y - 1$

k) $-9x - y + 1 + 5y + 5x - 10$

l) $-x + 8 + 6x - 4y - 8x + 3$

m) $4x - 7 + y - 7x - (-3y)$

n) $8x - 5y - x + 9 - y$

6 Simplify each of these expressions.

   a) $4a^2 + 3a + 7 - a^2 - 2a - 8$

   b) $a^2 + b^2 - a^2 + b^2$

   c) $x^3 + 7x^2 - x - 3x^2 + 2$

   d) $4xy + 2yx - 8xy + 3$

   e) $2a - 3ab + 2ba - b$

   f) $7t + 4s - 8t - 9s$

   g) $7yz + 3z - 2zy - 2z$

   h) $a^2 - b^2 + a - b + ab - ba$

   i) $12pqr - 2abc + 7pqr - bca$

   j) $3a^2b + 2ab^2 - 2a^2b - 2ab^2$

   k) $7abc - 4bca + 6cba - cba$

   l) $5ab - 1\frac{1}{2}\,ab + \frac{3}{4}\,bc + \frac{1}{4}\,cb$

   m) $\frac{1}{2}x + \frac{1}{2}y - \frac{1}{4}x + \frac{2}{3}z + y$

7 Simplify each of these expressions as much as you can.

   a) $(3c + a) + (2b - c) + (a + b + c)$

   b) $(z - x) + (y - z) + (x + y)$

   c) $(5a^2 + 7c^2) + (2c^2 - b^2 + a^2) + (a^2 + b^2 - c^2)$

   d) $(b^2 - k^2) + (-5b^2 - 7k^2) + (2b^2 + 4k^2)$

   e) $(2cb + 5ab) + (5ac - 5ba) + (2ab + 3bc)$

   f) $(a^3 - b^2) + (5b^3 + 2a^2 + 5c) + (2a^3 + 3b^2 - c)$

   g) $(5abc - 7cb + 4ac) + (4cba - 4bc + 3ca)$

   h) $4ab + (6bc + 5a^2b - 2ab^2) + (9ab^2 - 7ab)$

   i) $(8x^2 + 9x^3 - 4) + (4x^2 + 3x - 1) + (8x^3 + 7x^2)$

   j) $\left(\frac{1}{2}x^2yz - xy^2z\right) + \left(x^2yz + xy^2z - xyz^2\right)$

   k) $\left(\frac{1}{6}xy^2 + xy\right) + \left(\frac{1}{3}xy^2 - \frac{1}{4}yx\right) + \left(\frac{1}{2}xy\right)$

   l) $\left(\frac{2}{3}yz - \frac{1}{4}xyz - xz\right) + \left(\frac{3}{4}xyz - \frac{1}{2}xz\right)$

   m) $(7x^2 + 5x^3 - 4x - 5) - (2x^3 + 5x^2 - 4x + 3)$

   n) $(14 + 2x - 12x^2 + 7x^3) - (5x^3 + 4x^2 - 3x - 2)$

   o) $(2x - 5x^3 + 4) + (2x^3 - 5x + 7) - (2x^2 + 3x^3 - 5)$

   p) $(6x^3 - 2x + 10x^2) - [(2x - 5x^3 + 1) + (5x^2 - 7x + 4)]$

   q) $(3x^2 - 8x) + (15 - 4x - 3x^3) - [(5x - 4x^3) + (2x^2 - 7x + 4)]$

8 Simplify each of these expressions as much as you can.

a) $2m \times (-7n)$

b) $3a \times 5b$

c) $-4b \times (-8b)$

d) $2k \times (-7k)$

e) $-\frac{3}{4} y \times \left(-\frac{8}{9} y\right)$

f) $-\frac{1}{2} x \times 6x$

g) $-\frac{3}{5} u \times (-20v)$

h) $54b \div 9a$

i) $(-24m) \div (-18n)$

j) $3a \times 6a \times (-4b)$

k) $2xy \div 3y^2 \times 5x^2$

l) $\frac{c^2}{5} \div \frac{cd}{25}$

m) $5d \times 3de \times 2def$

n) $4ab \times 3bc \div (-9a^2c)$

o) $\dfrac{8gf^2 \times (-2g^2f)}{12fg \times (-f^2)}$

9 Expand and simplify each of these expressions as much as you can.

a) $4(m - 9)$

b) $14(f - 3)$

c) $15(k - 2m)$

d) $2(3x + 4)$

e) $6(2p + 3q)$

f) $5(3d - 2e)$

g) $-(-x - y)$

h) $-(-x + y)$

i) $-(b - 3)$

j) $-3(3r + s)$

k) $-2(2m - 3n)$

l) $-9(2p - 6)$

m) $-3(-r + 2s)$

n) $3x - 2(y - x)$

o) $8(2x - 3) - 6x$

p) $9 - 2(4m + 5)$

10 Expand and simplify each of these expressions as much as you can.

    a) $-7a - 3(9 - 7a)$

    b) $7 - 4b + 2(5b + 8)$

    c) $-3(4c - 1) + 9c + 8c$

    d) $10 - (2e - 6) - e$

    e) $4(3 + 7f) + 6(2 - f)$

    f) $9(2g - 4) - 2(7g - 12)$

    g) $5(-3h - 1) - (6 - 5h)$

    h) $2(7 + 6k) + 15(-1 + k)$

    i) $(-9m + 5) - 8(-m - 2)$

    j) $-(5n - 6) + 4(3 + 5n)$

    k) $3(p - 2q) - 2(3p - q) + 6(p - q)$

    l) $(r + s) - 2(5r - s) - 4(3r - 2s)$

    m) $2(3t + u) - 5[3(t - 3u) - 4(2t - u)]$

    n) $7[(v - w) - 2(v + 3w)] - 4(v - 13w)$

    o) $2x - 3\{2(5x - y) - 4[x - (7x - y)]\}$

    p) $[2(y + 5z) - 3(y - z)] - 7[3y - (y + 6z)]$

11 Expand and simplify each of these expressions as much as you can.

    a) $x(x + 1) + 3(x + 2)$

    b) $a(a - 4) + 2(a + 3)$

    c) $b(b + 1) - 5(b + 1)$

    d) $2c(2c - 5) + 5(2c - 5)$

    e) $d(d - 3) - 6(d - 3)$

    f) $e(5 - e) - 6(2 + e)$

    g) $3x + 2x(5x - 7)$

    h) $9x - 3x(2x - 4)$

    i) $8x - 6x(3 - 2x)$

    j) $-5x + 5x(x + 4)$

    k) $4n(6n + 9) - 10n^2$

    l) $14n - 3n(4n - 1)$

    m) $-8n^2 - 8n(-4 - 2n)$

    n) $7k^2 - 2k(3k + 1) - 9k$

    o) $-6k + 5k(8 - k) - 8k^2$

    p) $k^2 + 1 - 4k(2k - 9)$

    q) $-10k^2 - 3k + 2k(5 + 6k)$

    r) $8y + 9y^2 + 4y(11 - 2y)$

    s) $-4y(-2y - 7) + 6y^2 - 7y$

12 Expand and simplify each of these expressions as much as you can.

a) $9y - 3y(-4 + 3y) + 12y^2$

b) $5y(2y - 4) + 2y(y + 9)$

c) $-4u(3u - 1) + 7u(3 - 2u)$

d) $6u(-5u + 1) - 3u(4u - 12)$

e) $3u(-u - 5) + 8u(2u + 1)$

f) $7x(2x - 4) - x(10 - 3x)$

g) $-6x(5 + x) + x(13x + 1)$

h) $-t(12 - 4t) + 8t(10 - t)$

i) $7t(2t + 2) - 9t(-1 + 6t)$

j) $4p(-3p - 5) - p(10 + 4p)$

k) $6t^2 + t(14t - 5) + t(17 - 3t)$

l) $-t(2 - t) - 3t(6 + 8t) - 12t$

m) $b(4b - 9) + 8b(2b + 3) - 7b^2$

n) $11b - b(3b - 6) + 2b(4b + 5) - b^2$

o) $-3b(b + 2) + 2b(b - 2) - 12b^2$

p) $5n(2n^2 + n) - 3n(8n^2 - 2n)$

q) $n^2(4n - 3) - 6n^2(4n - 9)$

r) $-2a^2(9 - a - 4a^2) + 4a^2(a^2 - 2a + 3)$

1 Write down the size of the complement of each of these angles.

    a) $9°$

    b) $67°$

    c) $52°$

    d) $48°$

    e) $16°$

2 Write down the size of the supplement of each of these angles.

    a) $89°$

    b) $173°$

    c) $101°$

    d) $17°$

    e) $34°$

3 Use the rules of geometry to find the size of the angle *c* in each of these diagrams.

    a)

    b)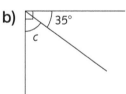

    c)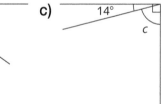

    d)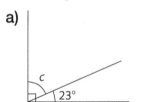

    e)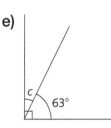

    f)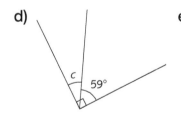

    g)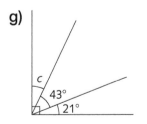

    h)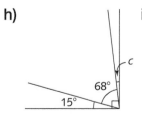

    i)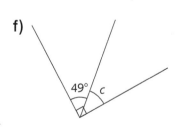

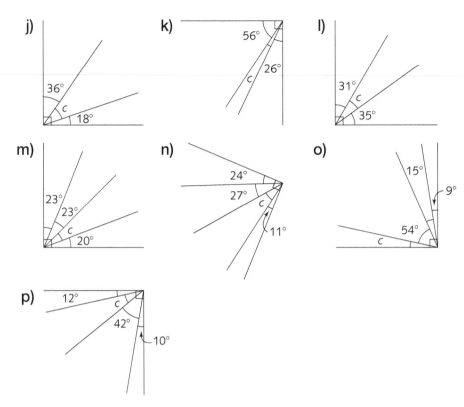

j) 36° c 18°

k) 56° 26° c

l) 31° c 35°

m) 23° 23° c 20°

n) 24° 27° c 11°

o) 15° 9° 54° c

p) 12° c 42° 10°

4 Use the rules of geometry to find the size of the angle s in each of these diagrams.

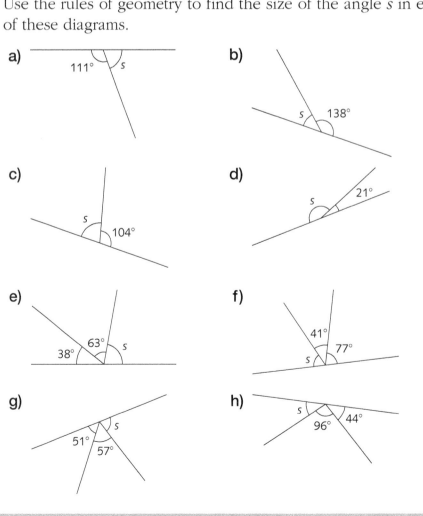

a) 111° s

b) s 138°

c) s 104°

d) s 21°

e) 38° 63° s

f) 41° 77° s

g) 51° s 57°

h) s 96° 44°

i)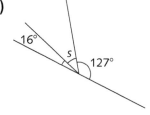

j)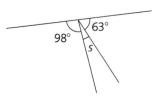

k)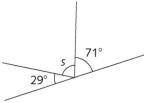

l)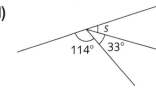

5 Use the rules of algebra and geometry to calculate the value of *a* in each of these diagrams.

a)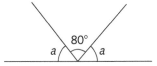

b)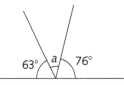

c)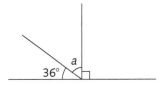

d)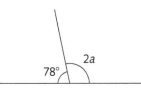

e)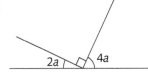

f)

g)

h)

i)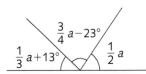

j)

k)

l)

6 Use the rules of algebra and geometry to calculate the values of g and h in each of these diagrams.

a)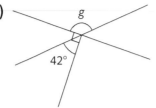

b)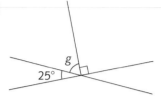

c)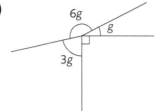

d)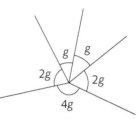

e)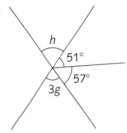

f)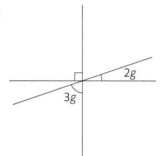

g)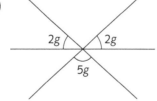

h)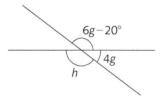

i)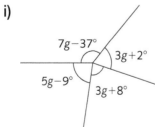

j)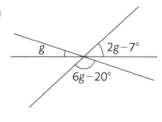

k)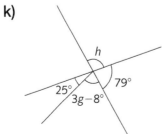

l)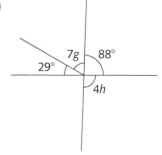

7 Use a ruler and a pair of compasses only to construct an accurate drawing of $\triangle ABC$ with $AB = 6$ cm, $BC = 4$ cm and $AC = 5$ cm.

8 Use a protractor and a ruler only to construct an accurate drawing of $\triangle ABC$ with $AB = 5$ cm, $\angle CAB = 30°$ and $\angle CBA = 50°$.

9 Use compasses, protractor and a ruler only to construct an accurate drawing of $\triangle ABC$ with $AB = 6$ cm, $AC = 4$ cm and $\angle ABC = 40°$.

Now use a protractor to measure $\angle CAB$ and $\angle ACB$.

10 Use compasses and a ruler only (no protractor) to construct an accurate drawing of $\triangle ABC$ with $AB = 8$ cm, $BC = 5.5$ cm and $\angle ABC = 45°$.

Now use a protractor to measure $\angle CAB$ and $\angle ACB$.

11 Use compasses and a ruler only (no protractor) to construct an accurate drawing of rectangle $ABCD$ with $AB = 5$ cm and $BC = 3$ cm.

12 Use compasses and a ruler only (no protractor) to construct an accurate drawing of square $ABCD$ with sides of 5.5 cm.

13 Here is a rough sketch of a geometric shape.

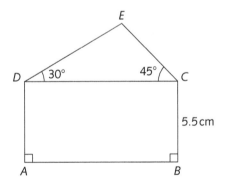

a) Use compasses and a ruler only (no protractor) to construct an accurate drawing of this shape so that
$$\frac{1}{2}\,AB = AD = BC = 5.5\text{ cm}$$

b) Use a ruler to measure the lengths of DE and EC on your diagram.

14 Here is a rough sketch of a geometric shape.

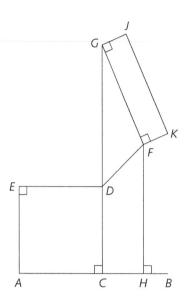

a) Use compasses and a ruler only (no protractor) to construct an accurate drawing of this shape so that

- $AB = 13.9$ cm
- $AC = CB = CD = DE = AE$
- $DF = 5$ cm, $DG = 11$ cm and $GJ = 2$ cm
- $GJ = FK$
- $\angle GDF = 45°$

b) i) Use a ruler to measure the lengths of CH, HB, FG and FH.

ii) Use a protractor to measure the sizes of $\angle DGF$ and $\angle DFG$.

An introduction to coordinate geometry

1 On a piece of graph paper, draw and label the x- and y-axes to make a Cartesian plane.

Now plot (and label) each of these points.

$A(1, 2)$	$B(-7, 0)$
$C(0, -5)$	$D(-1, 1)$
$E(7, 7)$	$F(0, -4)$
$G(7, 0)$	$H(1, 6)$
$I(-8, -2)$	$J(6, 3)$
$K(-6, -6)$	$L(2, -4)$
$M(3, -3)$	$N(-1, 4)$
$O(-3, 6)$	$P(4, 7)$
$Q(3, 5)$	$R(-5, 5)$
$S(3, 3)$	$T(-7, 2)$
$U(6, -1)$	$V(0, 3)$
$W(-3, -4)$	$X(1, -6)$
$Y(5, 5)$	$Z(-4, 4)$

2 Write down the x-coordinate of each of these points.

a) $(1, 0)$ **b)** $(2, 1)$

c) $(-4, -3)$ **d)** $(3, -2)$

e) $(-2, 3)$ **f)** $\left(0, -4\frac{1}{2}\right)$

g) $\left(-2\frac{1}{2}, 5\right)$ **h)** $\left(1\frac{1}{2}, -1\right)$

3 Write down the y-coordinate of each of these points.

a) $(-2, 4)$ **b)** $(-3, -1)$

c) $(1, -2)$ **d)** $(-4, 0)$

e) $(0, -3)$ **f)** $\left(2\frac{1}{2}, -3\frac{1}{2}\right)$

g) $\left(-6, \frac{1}{2}\right)$ **h)** $\left(-5, 1\frac{1}{2}\right)$

4 Write down the coordinates of each of the points *A* to *L* marked on this Cartesian plane.

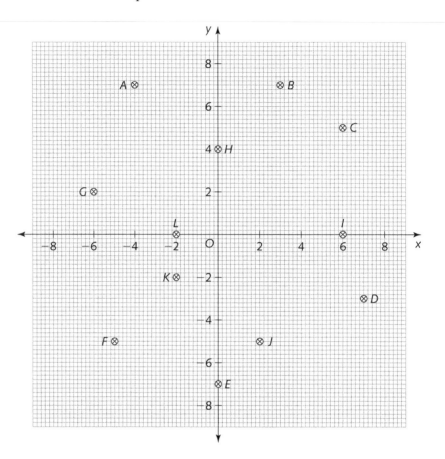

5 a) On a piece of graph paper, draw and label the *x*- and *y*-axes to make a Cartesian plane.

Now plot (and label) each of these points.

A (6, 12)	*B* (6, 8)
C (4, 6)	*D* (0, 4)
E (2, 3)	*F* (2, 1)
G (0, 0)	*H* (1, −1)
I (18, 18)	*J* (26, 22)
K (32, 22)	*L* (32, 14)
M (26, 0)	*N* (14, −2)
P (14, −4)	*Q* (24, −6)
R (28, −8)	*S* (28, −11)
T (25, −12)	*U* (25, −15)
V (22, −15)	*W* (22, −17)
X (18, −18)	*Y* (16, −19)

$Z(14, -19)$ $A^*(14, -20)$

$B^*(11, -20)$ $C^*(2, -10)$

$D^*(2, -16)$ $E^*(1, -19)$

$F^*(0, -20)$ $G^*(-1, -19)$

$H^*(-2, -16)$ $I^*(-2, -10)$

$J^*(-11, -20)$ $K^*(-14, -20)$

$L^*(-14, -19)$ $M^*(-16, -19)$

$N^*(-18, -18)$ $P^*(-22, -17)$

$Q^*(-22, -15)$ $R^*(-25, -15)$

$S^*(-25, -12)$ $T^*(-28, -11)$

$U^*(-28, -8)$ $V^*(-24, -6)$

$W^*(-14, -4)$ $X^*(-14, -2)$

$Y^*(-26, 0)$ $Z^*(-32, 14)$

$A^{**}(-32, 22)$ $B^{**}(-26, 22)$

$C^{**}(-18, 18)$ $D^{**}(-1, -1)$

$E^{**}(0, 0)$ $F^{**}(-2, 1)$

$G^{**}(-2, 3)$ $H^{**}(0, 4)$

$I^{**}(-4, 6)$ $J^{**}(-6, 8)$

$K^{**}(-6, 12)$

b) Join the points in alphabetical order, with straight lines. (Join A to B, B to C, C to D, . . . and so on. Finish by joining J^{**} to K^{**}.)

What picture have you drawn?

6 Write down the coordinates of each of the vertex points (corners) of these geometric shapes.

a)

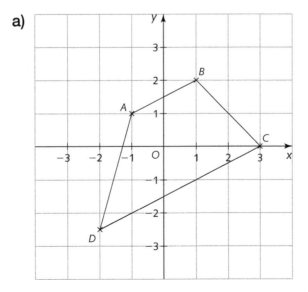

b)

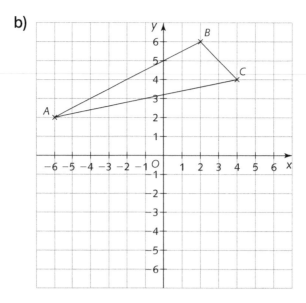

c)

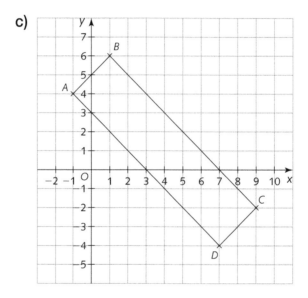

d)

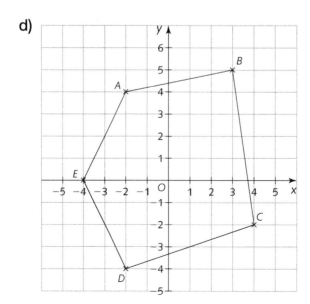

e)

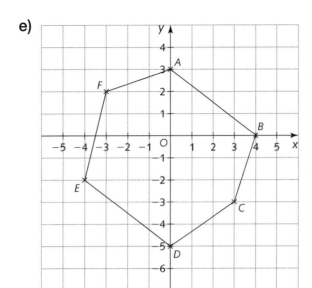

f)

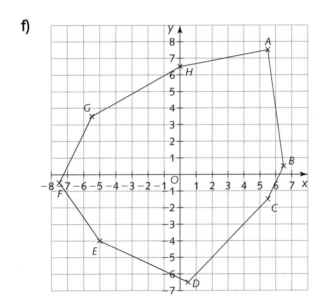

7 a) Plot the points $A(3, -5)$, $B(-3, -5)$ and $C(-3, 4)$ on a Cartesian plane.

  b) Plot the point D so that $ABCD$ is a rectangle.

  c) Draw the rectangle and write down the coordinates of point D.

8 a) Plot the points $A(-4, 7)$, $B(3, 7)$ and $C(3, -3)$ on a Cartesian plane.

  b) Plot the point D so that $ABCD$ is a rectangle.

  c) Draw the rectangle and write down the coordinates of point D.

9 a) Plot the points $A\left(0, 4\frac{1}{2}\right)$, $B\left(-3, 2\frac{1}{2}\right)$ and $C\left(1, -3\frac{1}{2}\right)$ on a Cartesian plane.

b) Plot the point D so that $ABCD$ is a rectangle.

c) Draw the rectangle and write down the coordinates of point D.

10 a) Plot the points $A(-2, -3)$, $B(3, -1)$ and $C(1, 4)$ on a Cartesian plane.

b) Plot the point D so that $ABCD$ is a square.

c) Draw the square and write down the coordinates of point D.